Meeting America's Security Challenges Beyond Iraq

A Conference Report

Sarah Harting

Prepared for the United States Air Force

PROJECT AIR FORCE

ANALYSIS & SOLUTIONS

Publication of this report was sponsored by the United States Air Force under Contract FA7014-06-C-0001. Further information may be obtained from the Strategic Planning Division, Directorate of Plans, Hq USAF.

Library of Congress Cataloging-in-Publication Data

Harting, Sarah.
    Meeting America's security challenges beyond Iraq : a conference report / Sarah Harting.
      p. cm.
    Includes bibliographical references.
    ISBN 978-0-8330-4469-3 (pbk. : alk. paper)
      1. National security—United States—21st century—Congresses. 2. United States—Military policy—21st century—Congresses. 3. United States—Armed Forces—Organization—Congresses. 4. United States. Dept. of Defense—Appropriations and expenditures—Congresses. I. Title.

UA23.H3668 2008
355'.033073—dc22

2008028846

The RAND Corporation is a nonprofit research organization providing objective analysis and effective solutions that address the challenges facing the public and private sectors around the world. RAND's publications do not necessarily reflect the opinions of its research clients and sponsors.

RAND® is a registered trademark.

Published 2008 by the RAND Corporation
1776 Main Street, P.O. Box 2138, Santa Monica, CA 90407-2138
1200 South Hayes Street, Arlington, VA 22202-5050
4570 Fifth Avenue, Suite 600, Pittsburgh, PA 15213-2665
RAND URL: http://www.rand.org/
To order RAND documents or to obtain additional information, contact
Distribution Services: Telephone: (310) 451-7002;
Fax: (310) 451-6915; Email: order@rand.org

Preface

On December 4, 2007, the RAND Corporation and the Center for Naval Analyses (CNA) Corporation cohosted a conference entitled, "Meeting America's Security Challenges Beyond Iraq." The conference, which included approximately 70 senior analysts from selected think tanks and academic institutions, high-ranking officers from the military services, selected government officials, and several journalists, was held at the offices of the RAND Corporation in Arlington, Virginia. The purpose of this conference was to provide leading analysts and practitioners of U.S. security policy and strategy with an opportunity to assess key challenges in the emerging security environment and to consider implications for strategy, force posture, and investment priorities.

This conference report provides an overview of the major topics, themes, and issues addressed at the conference as interpreted by RAND and CNA staff. To encourage a candid discussion of issues, the conference was held on a "not for attribution" basis; hence, the identities of the presenters and discussants are not provided.

RAND Project AIR FORCE

RAND Project AIR FORCE (PAF), a division of the RAND Corporation, is the U.S. Air Force's federally funded research and development center for studies and analyses. PAF provides the Air Force with independent analyses of policy alternatives affecting the development, employment, combat readiness, and support of current and future aerospace forces. Research is conducted in four programs: Aerospace Force Development; Manpower, Personnel, and Training; Resource Management; and Strategy and Doctrine. Additional information about PAF is available on our Web site:
http://www.rand.org/paf/

Center for Naval Analyses

The Center for Naval Analyses, a division of the CNA Corporation, is the U.S. Navy's federally funded research and development center for studies and analyses. The Center for Naval Analyses provides support to the Navy and Marines across the spectrum of defense activities, from personnel to technology, to operations and readiness. It does this by working directly with operating forces, as well as Navy, Marine Corps, and joint commands, in war and in peace. For over 60 years, since their groundbreaking work with the Navy during World War II,

CNA's goal has been to use scientific techniques to support the effective use of naval forces—and other defense concerns. Additional information about CNA is available at http://www.cna.org

Contents

Preface iii

Summary vii

Acknowledgments xi

Abbreviations xiii

Introduction and Overview 1

SESSION I

Overview of the Emerging Security Environment 3

Five Trends Affecting U.S. Security 3

"Five Long Wars": America's Security Landscape Beyond Iraq 4

Threads of Instability 5

Question and Answer Session 5

SESSION II

Key Threats and Challenges for the Future 7

China as a Strategic Challenge 7

Nuclear-Armed Regional Adversaries 8

Question and Answer Session—Part 1 9

Terrorism and Insurgency: The Changing Threat 9

Question and Answer Session—Part 2 10

Iran: A Multidimensional Challenge 10

Question and Answer Session—Part 3 11

Midday Presentation 13

Question and Answer Session 14

SESSION III

Priority Capabilities for Securing U.S. Interests 15

Achieving Global Effects 15

Defeating "Hybrid Threats" in the 21st Century 15

Capabilities Required for Countering State Adversaries 16

Understanding Possibilities and Risks 16

Question and Answer Session 17

SESSION IV

Fielding the Requisite Forces in a Resource-Constrained Environment 19

Federal Budget Trends and the Outlook for Defense Programs 19

A New Division of Labor Among America's Armed Forces 20

U.S. Ground Forces: Options for Future Development 20

Question and Answer Session 21

Closing Remarks 23

Question and Answer Session 24

Conference Agenda 25

Bibliography 27

Summary

On December 4, 2007, RAND Project AIR FORCE and the Center for Naval Analyses hosted a conference, "Meeting America's Security Challenges Beyond Iraq." From the outset, it was clear that the phrase "beyond Iraq" does not mean that supporting the emergence of a stable Iraq will soon disappear as a major security issue for the United States. On the contrary, all participants recognized that the wars in Iraq and Afghanistan are likely to persist well into the future. The security obligations associated with these conflicts would join the list of ongoing security obligations the United States must already address, such as countering terrorists, deterring Iran and keeping open the Straits of Hormuz, promoting peace on the Korean peninsula and across the Taiwan Strait, and strengthening alliances more generally.

There was broad consensus throughout the discussion that the following are among the issues and challenges to U.S. foreign and security policy that any new administration will have to face:

1. The United States needs to restore its credibility as a world leader in the face of widespread anti-American sentiments. Regaining an influential voice in regional security matters was also considered necessary for protecting and advancing U.S. and allied interests. (See pp. 3–5.)
2. The executive branch needs to greatly improve the integration of interagency approaches to complex security problems. This is particularly urgent because the problems that arise when combining traditional and irregular warfare, as in Iraq and Afghanistan, have historically persisted for many years and will likely continue in the indefinite future. (See pp. 4–6.)
3. Dealing with the emergence of China in all its many manifestations is going to be a multigenerational issue for the United States. In the security field, it will be important that, as China improves its military capabilities, the United States maintain key advantages or risk losing its credibility as a guarantor of stability in East Asia. (See pp. 3–4.)
4. U.S. defense planners need to prepare to counter nuclear-armed regional adversaries in case nonmilitary activities fail to prevent the proliferation of nuclear weapons. Aside from the longstanding concern about the possibility of proliferation to terrorists, the reality of nuclear-armed "rogue regimes," such as North Korea and, potentially, Iran, could profoundly alter regional security environments in ways that would be extraordinarily harmful to U.S. interests. (See p. 4.)
5. The nexus of rapid population growth in the Islamic world, stalled economic growth, and vehement anti-Americanism suggests that the "long war" will indeed be long. The

so-called youth bulge creates huge potential recruitment pools for jihadist extremists. (See pp. 3–4.)

While these concerns reflect a broad consensus among the conferees, there was less agreement about how to respond. If the United States is to meet these challenges, along with existing security obligations, its armed forces will be called upon to undertake a range of "nontraditional" missions at a scale and level of intensity well beyond what today's forces can sustain.

In turn, conferees discussed whether the notion of complex or "hybrid" wars helps planners prepare for future conflicts. Some conferees linked the idea of hybrid wars with "regime change" operations and were convinced the Iraq experience has made it very unlikely that there will be any appetite in Washington for such endeavors in the future. Others argued that, given a long history of U.S. military action against regimes (going back to the Mexican War), it would be wrong to discount the possibility of future similar operations.

In addition, there was considerable discussion among conferees about whether U.S. ground forces need to be restructured or realigned to better conduct counterinsurgency, foreign internal defense, and constabulary missions. The debate revolved around competing visions of the future. On one hand, if Iraq and Afghanistan are going to be the final U.S. experience with counterinsurgency, restructuring forces might be unnecessary. On the other hand, if the United States is destined to be engaged in conflicts that include both traditional combat and counterinsurgency operations, some form of restructuring is merited. Several participants observed that the currently planned expansion of ground forces makes it easier to contemplate realignment.

Most conferees agreed that the Navy and Air Force should focus on major combat operations against regional powers to maintain the ability to employ overwhelming airpower with precision weapons from land bases and from the sea *in more than one region of the world*. Some saw this capability as an important "strategic hedge." By the same token, the contributions both services are providing to ongoing conflicts in Iraq and Afghanistan are ample illustration that the capabilities they bring to the fight are applicable across a wide array of circumstances today and will be into the future.

In short, there was widespread agreement that the set of the challenges the United States faces, coupled with the varied demands on the military forces, will require significant changes in how the Department of Defense trains, equips, and postures its forces and, more broadly, how the U.S. government is organized for the advancement of U.S. interests abroad. Prominent examples of such changes include the following:

- U.S. forces will need to be able to conduct, on a sustained basis, a large number of "train-equip-advise-assist" missions in countries striving to counter Islamist-based terrorist and insurgent movements. Substantial numbers of U.S. ground forces will be associated with this mission. (See pp. 15–16, 20–21.)
- Nonmilitary agencies, including the Department of State, the U.S. Agency for International Development, the Department of Justice, and the Intelligence Community, must play more visible and, ultimately, predominant roles in efforts to counter terrorist and insurgent groups abroad. (See pp. 9–10, 21, 23–24.)
- U.S. and allied forces will need to prepare for the possibility that regional adversaries, such as North Korea and Iran, may soon field nuclear weapons. This will raise a host of large, crosscutting policy issues (such as whether the goal of regime change through invasion

and occupation is still appropriate and feasible in the presence of a nuclear threat). It will also call attention to important capability shortfalls (such as the inability to locate and destroy nuclear weapons and their delivery means, as well as the ability to deploy effective, multilayered defenses against a missile attack with nuclear weapons). (See pp. 8–9, 13–14.)

- As China continues to modernize its conventional forces, U.S. planners will be compelled to rethink key aspects of the U.S. approach to power projection, as the U.S. posture in the Western Pacific is becoming increasingly vulnerable to Chinese antiaccess weapons. (See pp. 7–8, 16.)

Furthermore, while there was agreement that the interagency process sometimes borders on dysfunctional, there was little agreement on how to correct the problem. The next administration will have to determine, systematically and thoughtfully, how to address the budgetary, cultural, and planning differences among the departments of Defense, State, Treasury, and Agriculture and other parts of the government. It will also need to focus on timely implementation of policy. Today, it is almost impossible to reach agreement on how to define problems and identify solutions. And few agencies other than the Department of Defense have the capabilities necessary to conduct large-scale operations in austere environments—a growing need in many places.

Finally, pressures on defense spending within the overall U.S. budget cast a long shadow. The combination of existing requirements, including Iraq and Afghanistan, with future challenges suggests that the demand for defense resources will increase rather than decrease even as these conflicts subside. The ground forces need to recapitalize, as do the Air Force and Navy, as the inventories of frontline aircraft and ships reach and often exceed their expected service lives. No one expected defense budgets to remain at current levels, much less increase. All, however, acknowledged that difficult choices loom in the not too distant future.

Acknowledgments

We would like to thank several individuals at both RAND and CNA who provided key guidance and assistance throughout the preparation of this report. We extend our thanks to Andrew Hoehn, David Ochmanek, and Michael Spirtas at RAND; and Christine Fox, Michael McDevitt, and Daniel Whiteneck at CNA.

We also thank our project sponsors in the U.S. Air Force and the U.S. Navy. In particular, we acknowledge Maj Gen William Chambers, David Moniz, and Christy Nolta in the U.S. Air Force, in addition to the Naval Operations staff, for their support of this endeavor.

In addition, we are grateful to a number of RAND and CNA staff who helped to make this conference possible. In particular, Kathy Berens and Kathy Lewis at CNA; and Molly Coleman, Emily Taylor, Nathan Chandler, Leslie Thornton, Emily Daly, Katherine Mills, Matthew Lolich, and Leanna Ferguson at RAND.

Finally, we extend our sincere thanks to all our conference participants for their time and their willingness to express their views on these important policy issues.

Abbreviations

CNA	Center for Naval Analyses
DoD	Department of Defense
GDP	gross domestic product
MCO	major combat operation
PAF	Project AIR FORCE
PLA	People's Liberation Army
WMD	weapons of mass destruction

Introduction and Overview

The "Meeting America's Security Challenges Beyond Iraq" conference was organized into four main sessions:

- Overview of the Emerging Security Environment
- Key Threats and Challenges for the Future
- Priority Capabilities for Securing U.S. Interests
- Fielding the Requisite Forces in a Resource-Constrained Environment.

Each session featured several short presentations presided over by a panel moderator. Following these, conference participants offered questions and reactions to the presentations. (See Appendix A for the conference agenda.)

Following Sessions II and IV, a senior military officer provided remarks and participated in a question and answer session. The sections that follow provide brief summaries of all of the presentations along with the salient points of the ensuing discussions. To facilitate a frank and open exchange of ideas, the conference was conducted under Chatham House rules of nonattribution.

Overview of the Emerging Security Environment

How should the United States deal with trends? Shape them? Cope with them?
—Senior Military Analyst

The panel members in this session took a broad view of the defense challenges that the United States is likely to face over the next ten years and likely beyond. They also discussed a number of trends and long-term issues that will require the continued attention of U.S. military planners.

Five Trends Affecting U.S. Security

The first presentation made it clear that the phrase "beyond Iraq" should not be construed as suggesting that supporting the emergence of a stable Iraq will disappear as a major security issue for the United States any time soon. On the contrary, the presentation noted that the wars in Iraq and Afghanistan are likely to persist well into the future. The presenter argued that the security obligations associated with these conflicts would join the list of ongoing security obligations the United States must already address, such as countering terrorists, deterring Iran and keeping open the Straits of Hormuz, promoting peace on the Korean peninsula and across the Taiwan Strait, and strengthening alliances. These should be considered givens for thinking about the future.

In light of these obligations, the panelist focused on five trends that will affect the U.S. security environment. The first is the erosion of U.S. influence. This includes the perceived decline in U.S. power and credibility, which weakens U.S. efforts to deter adversaries, reassure allies, and shape regional security environments. The second is the rise of China as a global military and economic power, its capacity to translate that power into a coherent political and strategic challenge, and the widespread effects that China will have on the regional balance in Asia, which could negatively affect America's position as a guarantor of stability on the littoral of East Asia. The third trend was changes in demographics (and their likely effects on politics and security concerns in Europe, Russia, the Middle East, and Africa). Of particular concern is the so-called youth bulge in much of the Muslim world. The combination of a pervasive anti-American sentiment in many Muslim countries and a huge pool of potential jihadi recruits suggests that the "long war" will indeed be long. The fourth trend is the role of some energy-producing states (e.g., Russia, Iran, and Venezuela), which are being enriched by a cost of oil near US$100 per barrel and, as a result, will have the resources to pursue policies inimical to U.S. interests. This trend reflects the intersection of national security, energy economics, domestic politics, and the need for integrated solutions across government agencies and between the U.S. and its allies. The final trend discussed was the proliferation of nuclear weapons and related technologies. The concerns here are preventing the spread of such capabilities to nonstate adversaries, strengthening U.S. leadership of multilateral counterproliferation efforts, and reinforcing U.S. deterrence credibility. Of specific interest to long-range planners in the Department of Defense (DoD), the panelist concluded by outlining the potential strategic consequences of certain tactical military developments, such as antiship ballistic missiles and the global use of improvised explosive devices.

"Five Long Wars": America's Security Landscape Beyond Iraq

The second presentation provided a somewhat different perspective on five strategic challenges that the presenter argued will dominate America's future security landscape: radical Islam, China's growing power and security competition in Asia, regional powers with nuclear weapons, Russia's problematic resurgence, and shifting alliance and partner relationships.

The speaker asserted that radical Islam and the terrorist groups that it spawns will remain a challenge for many years to come. Successfully combating this challenge will require a shift from the approach the administration has taken since 2001. Specifically, instead of "traditional" warfighting capabilities, U.S. forces and other assets will need to conduct a sustained and sizeable effort to train, equip, advise, and assist host-nation forces to counter nonstate adversaries.

Secondly, the panelist discussed how China's growth fuels a strategic competition between Beijing and Washington that could be increasingly difficult to manage. This challenge is compounded by the fact that both Japan and India are emerging as well, although in different ways. Therefore, the United States will need to strengthen regional partnerships in the face of these challenges.

The panelist defined the third challenge—regional powers with nuclear weapons—in terms of a "weak" state seeking nuclear weapons as a deterrent against attack. It will be difficult for the United States to prevent such states from obtaining nuclear weapons and from considering using them in the event of a future conflict. Deterring regional nuclear powers during crises will be considerably different from deterring Russia and China. Meeting this challenge will require substantially improved active defenses and, perhaps, new approaches for power-projection operations. (See "Nuclear-Armed Regional Adversaries" in Session II.)

The fourth challenge, Russia's resurgence, arises from Russian economic growth coupled with the reestablishment of authoritarian governance and Moscow's willingness to assert itself in opposition to U.S. interests. These developments will likely continue and will require sustained attention from the United States.

Finally, the panelist focused on shifting alliances and partnerships, arguing that relations between the United States and its security partners are changing and deteriorating in many areas, partially because perceptions about the risks and rewards associated with being allied with the United States have been changing. These shifting relationships are fostering a new "strategic geometry" of sorts, as allies and partners recalculate costs and benefits.

In light of these challenges, the panelist noted that defense planning in the United States is becoming far more complex. While the capabilities for prevailing in major combat operations (MCOs) will still be essential, the nature of these conflicts is changing. At the same time, capabilities will be needed for other sorts of operations that are not "lesser included cases" of the MCO. Hence, resources and forces may not be fungible across this set of challenges. As a result, the speaker argued that the DoD must obtain and sustain awareness of these challenges, reconcile U.S. ends and means with those of its partners, and establish a new division of labor among the services to shape the unfolding challenges. (For more on this, see "A New Division of Labor Among America's Armed Forces" in Session IV.)

Threads of Instability

In response to the two presentations above, one panelist added that demographic trends; economic trends, such as globalization; competition for resources, such as petroleum; and climate change are testing the capacities of a number of states, increasing the potential for state failure. This, in turn, warrants greater concern for ungoverned territories. The panelist argued that all this is complicated by the fact that a number of traditional U.S. allies are facing demographic and fiscal challenges. Furthermore, these challenges necessitate an indirect approach, with the United States putting its emphasis on building partner capacity, and its allies putting theirs on strengthening internal sovereignty and tailoring their forces to be capable of working with the United States to address nontraditional threats.

The panelist then continued with a discussion of several specific challenges:

- "hybrid" forms of warfare[1]
- the use of weapons of mass destruction (WMD) outside the MCO context (e.g., threat reduction and consequence management)
- the potential collapse of a major ally
- the threat of a humanitarian disaster on a catastrophic scale.

The panelist argued that the United States needs to work on restoring its global influence and leadership position, because allies will continue to demand U.S. leadership on and engagement in traditional concerns and involvement in emerging areas of instability, from Africa to the Caucasus to Asia. The panelist added that addressing these challenges will require a high degree of U.S. engagement over sustained periods. The panelist also noted that many past planning assumptions are no longer valid and that the United States will need joint military teams to address different types of conflicts.

Question and Answer Session

The first comment focused on public opinion and, more specifically, on what can be done to foster positive perceptions of the United States internationally. The commentator mentioned the success of the model that was used during the Cold War and asked whether it would be

---

[1] In defining *hybrid challenges*, the panelist described them as overlapping those found in the 2006 Quadrennial Defense Review: irregular, disruptive, catastrophic, and traditional challenges. See U.S. Department of Defense, *Quadrennial Defense Review Report*, Washington, D.C., February 6, 2006, p. 19.

prudent to replicate such a model today. This participant argued that changing public opinion is not cheap, but lessons can be taken from past efforts.

Another participant said that efforts to address demographic and social trends of concern should focus on enhancing civil society in Islamic states.

Another attendee questioned how U.S. policies can target the underlying causes of some emerging challenges more effectively. Several panelists agreed that more needed to be done along these lines. For example, how can the United States marginalize support for extremists? How can the United States help reform corrupt political regimes? All these questions taken together formed a first impression of a key conference theme: the need for the U.S. government and DoD to pay attention to integrated interagency approaches to complex future security issues and the need to address the budgetary, cultural, and planning differences between the departments of Defense, State, Treasury, and Agriculture and other parts of the government in defining problems and identifying potential solutions.

A final question centered on the direction of the United States itself, both internally and externally. A conference attendee asked the panelists about the current U.S. federal budget deficit and what the United States should do to make itself more financially solvent. One panelist responded by stating that meeting the challenges of the emerging security environment will require the United States to maintain a level of military spending somewhat higher (in terms of percentage of gross domestic product [GDP]) than what it had spent in the immediate aftermath of the Cold War.

Key Threats and Challenges for the Future

Deterrence alone is not a fully satisfactory answer to nuclear-armed regional adversaries.

—Senior Defense Analyst

Victory over terrorists and insurgents will come not from 'capture and kill' operations, but by breaking the cycle of radicalization, recruitment, and training. . . . We need to focus on ways to shape the perceptions and options of those who are children today in the Islamic world.

—Senior Defense Analyst

While the first panel took a global view of emerging security challenges, the second focused on a few distinct threats and challenges. Specifically, panel members focused on four main topics: China, nuclear-armed regional adversaries, terrorism and insurgency, and Iran.

China as a Strategic Challenge

The first speaker discussed key dimensions of the strategic challenges China poses. The speaker argued that China's reemergence is one of the most significant political developments of our time. China is in the midst of a drive to modernize its military forces. The panelist noted that the country can now shape the environment rather than just react to it. Two factors of particular note in China's reemergence are the speed at which this is taking place and how globalization is supporting the growth of the nation's economy and the spread of its influence.

The panelist argued that China's rise poses several challenges to international order, specifically with regard to economic and military issues. China's military and economic modernization is accompanied by political and diplomatic confidence exercised in regional and global forums ranging from the Six-Party Talks to the Shanghai Cooperation Organization and the United Nations. China's reemergence poses a number of unknowns: how China will use its economic clout; the abilities, agendas, and relationships among civilian elites in China; and whether the Communist Party will be able to transform itself to address its own range of domestic issues and concerns.

Next, the panelist questioned how the United States and its allies (e.g., the European Union, South Korea, Japan, and Australia) will react to a resurgent China. The panelist argued that concerns about the capabilities and institutional prerogatives of the People's Liberation Army (PLA) are warranted. The United States can assess the modernization of the PLA's forces with some confidence. This modernization effort is broad-based and ambitious and will profoundly affect the military balance in East Asia. For example, the PLA, long a ground-centric force, has now transformed into more of a naval- and air-centric force capable of protecting interests at home and abroad. The Chinese are applying both capabilities-based and contingency-based planning constructs. The panelist explained that, since the early 1990s, PLA efforts have focused on preparing for local wars. Recently, the PLA has shifted its attention to preparing for short but decisive conflicts and operations that depend heavily on command, control, communications, computers, intelligence, surveillance, and reconnaissance. It is also developing capabilities to deny U.S. forces access to the region. The panelist posited that U.S. forces will have to transform and modernize to "rise on the same tide" as the Chinese.

Turning to Chinese domestic issues, the panelist emphasized the increasing demographic bulge that is fostering a cadre of disenfranchised and potentially angry youth in China and noted that this bulge could have implications for China's internal stability. The panelist also questioned the degree to which China's modernization is addressing internal threats and, more specifically, the role the PLA is playing in addressing internal security. Furthermore, the panelist questioned the resilience of the PLA, noting that major investments in education and training will also be necessary if China is to succeed in transforming the PLA into a modern, effective fighting force.

Nuclear-Armed Regional Adversaries

The conference continued with a presentation on the nature of the threat that regional adversary states with the capability to use nuclear weapons pose. Such states as North Korea may be extremely weak in many respects, but their possession of nuclear weapons makes it possible that, despite U.S. military superiority, these states could severely constrain U.S. military options and operations in a confrontation with them. And, in fact, these adversaries might find they have incentives to use their nuclear weapons in war. Ironically, the political and military weaknesses of these states can make it difficult to deter them from escalating to nuclear use in a conflict.

The presenter noted that we should expect new nuclear states to probe to find the "red lines" of their adversaries. While superior U.S. military capabilities will have deterrent value against nuclear-armed regional adversaries, should the United States find itself in a conflict with such a state, deterring it from using one or more of its weapons could be problematic. These states have a number of options for using nuclear weapons, including test or demonstration detonations, high-altitude electromagnetic pulse shots, attacks against military targets, attacks against economic infrastructure, and attacks against urban areas. Attacks against targets at the lower end of the escalation ladder can cause disproportionate psychological effects. If the adversary threatens to attack urban economic targets and if the United States and its allies cannot prevent such attacks, the United States and its allies may be coerced into accepting some adversary demands. The presenter closed by noting that the current U.S. defense program will not provide capabilities sufficient to allow future commanders to have high confidence in operations aimed at *preventing* such adversaries from using their nuclear weapons.

In response to this presentation, one panelist stated that these states seek to avoid a direct conflict with the United States, which in turn provides us some leverage. That panelist also argued that these states face significant international political and diplomatic challenges to crossing the nuclear threshold, even under the most demanding scenarios. The presenter agreed and noted that steps can be made in peacetime to clarify U.S. "red lines." But the presenter added that if the U.S. finds itself at war with a nuclear-armed regional adversary, Washington will have to constrain its declaratory policies and military operations to limit the adversary's incentives to escalate.

Question and Answer Session—Part 1

Conference participants discussed the case of nonstate adversaries with nuclear weapons and agreed that this would constitute a grave problem, although a different one than the one state adversaries pose. Depending on the ideology and strategy of the group in question, a nonstate actor could be "undeterrable." Attendees also agreed that, when considering the decisionmaking of the actors in a nuclear conflict, the United States needs to account for the possibility of miscalculation. To meet this issue head on, the importance of reinvigorating the expertise across DoD and the government in nuclear-deterrence strategy and in regional expertise on potential adversaries was emphasized.

An attendee asked what steps the United States can take to deter Iran. A panelist responded by saying that there is no substitute for fielding adequate offensive and defensive military capabilities. Improved active defenses are a key means of making Iran's option to use nuclear weapons less attractive. Meeting participants also considered the importance of domestic political systems. For example, the Iranian regime draws on a wider base of support in its society than does the North Korean government. Therefore, Iran might behave more prudently than North Korea in a crisis or conflict. There was general agreement on the need for deterrence strategies and operations to be "tailored" to specific threats.

Terrorism and Insurgency: The Changing Threat

The speaker began by noting that the United States will have to address emerging threats while remaining engaged in Iraq and Afghanistan. He also argued that Salafist radicalization has only just begun; the enormous youth bulge in the Middle East has already created great difficulty in finding meaningful employment, and these youth may be disposed to adopt radical interpretations of Islam.

The presenter noted the emergence of a "cult of the insurgent." This image first arose following the 1967 Six Day War. Following the defeat of conventional Arab militaries by an Israeli force that used high technology to its advantage, the only significant force remaining in the Arab world was the Palestinian Fedayeen. This force turned to what are now known as irregular tactics to fight as a revolutionary vanguard.

The panelist then discussed what can be done to counter the threats posed by these types of adversaries. He said that the United States is generally good at killing and capturing people and is good at addressing immediate threats but is less adept in the strategic domain. It needs to turn its attention to breaking the cycle of recruitment and replenishment of terrorist adversaries. Such approaches argue for U.S. government investments in and coordinated uses of strategic communications and information operations. This approach would embed targeted kinetic operations within larger, long-term nonkinetic strategies.

In response, one panelist discussed how actions beget reactions and how religiously inspired revolts are often contagious. In light of the lessons of history and the advent of global, instantaneous communications, we should not be surprised to see widespread reactions to U.S. actions in Iraq and Afghanistan and to Israeli actions in Gaza, the West Bank, and Lebanon. The panelist argued that the United States faces significant difficulties in addressing how audiences in the Muslim world perceive U.S. actions. These difficulties are magnified by the weakness of modernizers in the Muslim world and U.S. ties to unpopular authoritarian Arab governments.

Question and Answer Session—Part 2

Conference participants discussed the extent to which Iraq is likely to remain a breeding ground and training center for radical jihadist elements. One panelist discussed how different aspects of state repression, such as timing and comprehensiveness, affect resistance movements. Participants also discussed a report released earlier in the year that showed the degree to which Muslims accept or reject political violence undertaken by radical Islamist groups.[1] Attendees discussed how such reports address questions of identity and whether people think of themselves as being Muslim, French, or Muslim and French, for example.

Another attendee observed that the panelists seemed to use the terms *terrorism* and *insurgency* synonymously. This attendee argued that terrorism should be regarded as a subset of insurgency. The larger threat, therefore, should be addressed overall as an insurgency. In response, a panelist noted that the perceived conflation of the two terms was not intentional. The panelist argued that the essence of the insurgency is competition for public support, while terrorists see themselves as revolutionary vanguards; in a sense, terrorists dream of becoming insurgents.

Iran: A Multidimensional Challenge

The next presentation considered the limits and reach of Iranian power. In an examination of Iranian strategic culture and how Iranians perceive threats and opportunities, the panelist argued that Iranian perceptions of their security needs largely transcend factional differences within the country. Some of Iran's primary objectives include defense of the homeland, support of proxies as a form of "forward defense," and a shifting of focus toward the East to engage key countries in Asia as part of an effort to offset pressure from the West.

The panelist argued that, to some extent, the United States helped Iran by removing adversarial regimes in Iraq and Afghanistan. However, Iranian interests in how each of these countries evolves have led it to play active roles in both places, often in opposition to U.S. policy preferences. Iranian activism has come at a price. There is evidence of a debate within the regime about the economic costs of Iran's antagonistic policies vis-à-vis the West and the isolation these policies have engendered.

Next, the panelist discussed a number of key themes—limitations, liabilities, and risks—and noted that there is a gap between the aspirations and reality of Iranian conventional mili-

---

[1]   Pew Research Center, *Muslim Americans: Middle Class and Mostly Mainstream*, Washington, D.C., May 22, 2007.

tary capability. The panelist argued that Iran's defense strategy is one of massive mobilization and societal resistance, based on the belief that ideological fortitude will enable Iran to outlast any opponent. Iran has sought to increase the ability of its forces to conduct joint operations, but there is little or no evidence that they have made significant progress in this regard. The presenter also asserted that it should not be assumed that Iranian "proxies," such as Hezbollah and Hamas, will always act in support of Iranian interests, despite the substantial support that Tehran provides such groups.

Another panelist responded to this presentation by pointing out that, if the recently released National Intelligence Estimate is accurate about Iran halting its nuclear weapon development effort,[2] it might be prudent to reconsider Iran's position of priority in U.S. defense planning. As a middle power without a nuclear weapon program, Iran has several inherent weaknesses that will hinder its ambition to expand its influence. First, its conventional military forces are weak. Second, there are sharp schisms within Iran's ruling elites, and recent tensions between Iran and the United States have heightened them. Furthermore, if the United States were to reduce its presence in Iraq, the panelist questioned whether Iran would be able to fill the void. The panelist argued that Iran would try, but that even the Shias within Iraq would resist this outside influence. Finally, the panelist noted that Iran has a number of economic problems, including the declining productivity of its oil fields, weakening domestic support for its ideological model, and international problems created by antagonistic diplomacy and harsh rhetoric.

Question and Answer Session—Part 3

One attendee asked the panelists how the recently released National Intelligence Estimate would affect perceptions about Iran's role in the region and whether the estimate would deflate that role. Another participant asked whether Iran was susceptible to an internal revolution, such as the one that took place in Ukraine. One panelist responded by noting that another revolution is unlikely in Iran, despite the fact that some in Iran may already perceive one to be occurring. The panelist argued that the reform movement in Iran is largely exhausted, that the regime will not hesitate to use violence to suppress any uprising, and that there is a tendency to underestimate the influence of the revolutionary guards.

---

[2] Office of the Director of National Intelligence, "Iran: Nuclear Intentions and Capabilities," *National Intelligence Estimate*, Washington, D.C.: National Intelligence Council, November 2007.

Midday Presentation

The DoD needs to change its model. We cannot look internally for all solutions; we have to go to our allies—go global, outsource, and look for capital and capacity. We lack the resources to be a global power on our own.

—Senior Military Officer

The midday presentation highlighted five issues that will continue to demand significant DoD attention. The first is globalization, which raises several questions for the armed forces of the United States. What are the strategic implications, and how can DoD best act and work in a globalized environment? Taking into account the declining importance of territorial boundaries and distances, how should DoD think about defense and its relations with other entities? How does the new media and strategic communications era affect how the United States fights? The speaker noted that the effects of globalization will be as far reaching as those of the transition from an agrarian to an industrial society. Given these implications, the speaker asked what DoD has done so far to address the changes globalization has caused.

The second issue is strategic uncertainty. At present, DoD does not know how such countries as Iran and Russia will evolve and to what extent they will seek to challenge U.S. interests in the future. The speaker asked how much of what these countries are doing is in reaction to U.S. actions and how much is motivated by their own agendas. And what, the speaker asked, should the U.S. policy in turn be? What steps can the United States take (or what tools are available for use) to reduce the amount of uncertainty in its relationships with these countries? For example, to what extent can the United States accept or, alternatively, limit the ability of Iran to export terrorism and extremist ideologies?

The third issue is the nexus between proliferation and WMD. U.S. forces have the ability and capabilities to see and track other maneuver elements, but not small entities, including individuals. U.S. forces are, in the speaker's judgment, "industrially biased" to build large things that can kill large things. To counter challenges like proliferation and WMD, U.S. forces will need to adapt and develop new capabilities to track and defeat "nontraditional" targets.

The fourth issue is U.S. relationships with its allies. The United States can no longer assume that its spending on research and development is adequate to keep pace with changes in the world or that its model of building and operating technology within closed-off organizations is capable of competing in a globalized, networked world. To remain competitive, the United States will need to work with its allies and will need to go "global" in its search for solutions to important security challenges. The United States lacks the resources and intellectual capital to be a global power by itself.

Finally, the fifth issue is the growing importance of the cyberspace domain and under-standing where this trend is leading.[1] The speaker argued that there is a blurring of the lines between nations and within such areas as research and development. The speaker expressed concern that cyberspace will be treated as analogous to space to the extent that space policy and programs will be classified and regulated in ways that inhibit others from participating. Instead, he argued that the cyber domain should be treated as an operating environment. The speaker argued that the National Security Agency has the capabilities and capacity to take a leading role in addressing issues about the cyber domain. The speaker also noted the need to integrate cyber operations into warfare. In the cyber world, information travels faster than the human mind's ability to make decisions. In such a world, machines will be doing the majority of the "fighting," and this raises serious questions regarding the ability of commanders at all levels to understand the situation and to formulate and execute appropriate strategies.

Question and Answer Session

The first question focused on how the United States should recruit people to work in the cyber domain. The speaker responded by calling for an increased interagency effort. The crosscutting nature of cyber activity requires efforts across bureaucratic boundaries, which leads to ques-tions about how to build such a decision process inside the U.S. government.

Another questioner focused on trends and challenges at the operational level and asked what capabilities are most needed to ensure that future U.S. forces will be able to prevail against adaptive adversaries. The speaker stated that conflicts will need "find and fix" tools—intelli-gence, surveillance, and reconnaissance capabilities—that are able to penetrate denied territory and survive. Such capabilities are also needed to conduct sensitive-site exploitation. Next, the speaker emphasized a need for diversification of "space tools"; such tools are extremely expen-sive, and current tools are not satisfactory. Third, the speaker emphasized, the need for cyber tools is increasing. He also said that the nonkinetic area has expanded to include all forms of soft power, but that U.S. strategy has not been broadened to make use of economic tools (for example, those at the disposal of the Department of the Treasury have not been fully lever-aged). The conclusion was that all these challenges require changes in DoD culture and that the leadership needs to remain on the cutting edge of institutional and process adaptation.

---

[1] The speaker reminded attendees that cyberspace—the *cyber domain*—includes both the networked computers and the communication links that tie them together.

Priority Capabilities for Securing U.S. Interests

We face an inherent security dilemma; China's strengthened defenses ultimately undermine our own.

—Senior Defense Analyst

Achieving Global Effects

This presenter began by stating that ends and means are at the heart of the challenge for achieving global effects. *Ends* are political goals that need to be translated into military objectives, while *means* are the tools that military commanders use to achieve the ends. The speaker pointed out several constraints on military operations that existed in the 1990s. First, the need to work with others complicated operations because of the need to achieve cohesion and overcome legal constraints. Time limits also shaped operations because there was little patience for extended peace operations. Third, there was an imperative on limiting civilian casualties, which affected target selection. Finally, efforts to avoid U.S. and allied losses constrained operations, influencing leaders to rely on air power.

The speaker noted that, more recently, acceptance of risk to forces and willingness to "go it alone" have increased. These two factors have contributed to a long-term commitment of ground forces, an increased number of U.S. casualties, and a greater hesitation from the international community to assist in operations. In the wake of Iraq, the speaker argued, a new model, utilizing the best attributes from both the 1990s and the post–9/11 environment, for conducting fully integrated operations (such as joint operations between air forces and special operations forces) may be more appealing for the future. Such a model could be very powerful, aligning political ends and military means without a long-term commitment such as those the United States now has in Iraq and Afghanistan. It would recognize the role of the forces outlined above, using the knowledge gained by forward presence and regional awareness, and would operate under clear political objectives and guidance.

Defeating "Hybrid Threats" in the 21st Century

The next speaker said that state-on-state wars are being displaced by hybrid wars and protracted wars. The adversaries in such hybrid conflicts have statelike capabilities, such as mobile missiles and antiarmor systems, and use irregular tactics.

The panelist noted that, even if Hezbollah, the Sunni insurgents in Iraq, and the Taliban in Afghanistan are representative of a significant portion of the future threat, U.S. capabilities for irregular warfare still need improvement. For example, military doctrine for irregular warfare now exists, but there is no comprehensive national-level approach. Furthermore, the capacity of U.S. intelligence systems is insufficient, as is the capacity for information operations and the supply of officers and noncommissioned officers qualified to serve as advisors to foreign forces. A robust irregular warfare strategy would emphasize indirect approaches, prevention, and discriminate responses. The speaker argued that the best force to carry out such a strategy is a "transformer force," or one capable of transforming to address the blurring of lines between conventional and irregular challenges. The panelist closed by stating that proposals for a division of labor that would create more distinct types of units optimized for maneuver warfare or irregular operations and the indirect approach are not warranted and may be inherently dangerous operationally and strategically.

Capabilities Required for Countering State Adversaries

This panelist began by stating that today's state-on-state challenges are more weighted toward air and maritime operations. He focused on the challenges China poses as the leading example of a modern, conventionally armed state adversary. China is building capabilities to deny access to its territory and areas of influence as a way of expanding its defense perimeter. China has shown increasing competence in a number of areas, including antisatellite and cyber attack capabilities, and is likely to target U.S. dependence on space and information. As China seeks to enhance its own security, it makes South Korea, Japan, and Taiwan feel less secure.

The panelist noted that the United States needs to continue to guarantee access to the commons in East Asia and to be capable of controlling the commons in a conflict.

Next, the panelist noted that the United States must reassure its friends and allies in East Asia. He recommended developing capabilities to defeat an antiaccess strategy. He also suggested that the United States develop such capabilities as theater missile defense; advanced networked forces; hardened or mobile bases; cyber defense; long-range strike capability; and more-versatile intelligence, surveillance, and reconnaissance assets.

Understanding Possibilities and Risks

In response, one panelist noted that each presentation dealt with the explosion of technologies and of risks. We are living in an extraordinarily unpredictable world, which poses problems for defense planners. He said that we also need to consider internal problems and that the U.S. defense establishment's strengths and weaknesses are intertwined.

The speaker said that one key attribute will be human capital. While there is strength in the identities of each of the military services, the U.S. military needs to develop officers that are more adaptable. One way to do this is to encourage career paths that cross service and departmental boundaries. It also needs to do a better job of integrating personnel from other departments into the planning and execution of national policy. This is a problem because almost everyone in the field of national security focuses on discrete problems, and few focus on broader issues that cut across lines of departmental responsibility.

Question and Answer Session

Conference attendees discussed whether hybrid conflict or asymmetric tactics are new. Some questioned whether Hezbollah is distinguishable from militia or guerilla forces that states have encountered for decades. One speaker noted that wars have always had political aspects and, usually, employed "asymmetric" tactics or strategies. Another panelist noted that one thing that has changed is that the United States has not yet seen a nonstate actor that has been capable of using state-level lethality.

Turning to implications for U.S. defense planning, one panelist recommended updating U.S. legal authorities to allow DoD more agility in countering threats. Another member of the audience argued for investing in flexible forces and leveraging the abilities of partners. Yet another discussed the idea that the Army might need to specialize, to designate units to deal with specialized problems such as stability and reconstruction, regime change, and WMD-capable adversaries.

There was consensus that it will be difficult to address many of these challenges with limited resources. Defense leaders will need to decide how best to adapt to respond appropriately with the means at their disposal. This will be a challenge for the next administration.

Fielding the Requisite Forces in a Resource-Constrained Environment

What should the division of labor be given resource constraints? Transform and adapt, or divide?

—Senior Military Officer

The last panel continued the theme of striving to meet new and adaptive challenges with limited means. The panel chair began by noting the ups and downs of U.S. budget cycles: About every 17 years, budgets tend to go down after prolonged hostilities. The chair also said that taxpayers want lower rather than higher taxes yet also want more personal benefits. This makes it inherently difficult to balance budgets.

Federal Budget Trends and the Outlook for Defense Programs

The first presenter gave an overview of federal budget trends and their implications for defense spending. He noted that the operations and support portion of the defense budget is on the rise, driven by increases in military pay and medical care. In addition, recent projections show that U.S. annual defense spending is expected to total approximately $521 billion in 2014.[1] The panelist also discussed past and projected spending for defense procurement. He argued that current procurement plans will not be able to sustain the force at its current size unless many platforms are allowed to reach unacceptable ages.

The panelist then discussed the level of spending in specific areas. The speaker noted that spending on shipbuilding is at or above Cold War levels. Investment projections indicate that spending is also roughly at Cold War levels for Air Force fighters and attack aircraft. However, the numbers of platforms being procured now at these funding levels are much smaller than during the Cold War.

Past and projected defense spending as a percentage of GDP showed that defense spending rose to 4 percent of GDP in fiscal year 2007 but is likely to decline.[2] In addition, the panelist noted that, absent changes in benefits, social security costs will rise in the long run to over 6 percent of GDP. Medicare and Medicaid costs will also exert pressure on the federal budget. The speaker also noted that DoD has publicized plans to field larger forces (specifically, ground forces), which will increase costs.

---

[1]   In current 2007 dollars. Congressional Budget Office, *Long-Term Implications of Current Defense Plans: Summary Update for Fiscal Year 2008*, Washington, D.C., December 2007, p. 2.

[2]   Congressional Budget Office, 2007, p. 4.

A New Division of Labor Among America's Armed Forces

The speaker began by noting that the United States does not have the ability to pick and choose its challenges. It appears that the period the United States is entering is more difficult and more challenging than any in the recent past. At the same time, the United States will continue to face tight resource constraints. To respond adequately, the military services might need to take on more differentiated roles.

The panelist challenged the notion of a "spectrum of conflict," noting that the nation's security challenges are distinct and that capabilities suited for one type of problem may not be applicable to another. To counter terrorists and insurgents, ground forces should focus more on training, advising, and assisting foreign partners. To counter state adversaries having nuclear weapons, air and naval forces should concentrate on prevention and denial. To counter an emerging peer, the United States must retain key power-projection advantages. In an environment of constrained resources, addressing these disparate challenges effectively may require some role specialization among and within the services.

In response to these remarks, one panelist said that he had never seen as much interservice rivalry as there is today. The next Secretary of Defense will have to find a way to ameliorate interservice competition and, simultaneously, forge better relations between DoD and other government departments. He supported the idea the previous speaker had raised about promoting some specialization in the capabilities the services field as a means of reducing costs. He said that many in the defense community overestimate the extent to which allies can share the burden of defense because they have different perceptions of the nature of the challenges the United States faces. He agreed that we are likely to face stark fiscal choices in the future but will also have to respond to the possibility of a protracted effort against terrorists and insurgent adversaries and a potential conflict with China. Moreover, the United States should realize the importance of new security challenges brought about by global climate change.

U.S. Ground Forces: Options for Future Development

This speaker questioned the idea that the United States will be able to reduce defense spending in the next few years, pointing out the likelihood that substantial numbers of U.S. ground forces are likely to remain committed in Iraq and Afghanistan. These extended deployments have placed considerable strains on ground forces that are likely to ease only marginally.

He argued that the Army is the key force in prosecuting the long war. It is the only service with the capacity to handle the long-term land engagements required in this conflict. To respond adequately, the Army will need to become a more-senior force at both the officer and noncommissioned officer levels. It will also have to make some choices about what type of force it will be.

The presenter continued by arguing that there is a compelling case for making constabulary missions the primary role for large parts of the Army. This will be a highly important issue for the next administration. The Army will also have to make choices about its future global force posture and force rotation policies if it is to accommodate the need to post troops in different theaters for extended periods.

In response to this presentation, one panelist returned to the idea of hybrid threats. To respond to these threats, he argued, future operations will have more to do with outreach, such as partnership building and fostering interoperability. The panelist discussed a number of tasks associated with countering hybrid warfare tasks, including defeating terrorists and insurgents, securing routes for convoys, training national armies and police forces, providing relief and logistical support to populations, supporting reconstruction operations, command and control efforts (including interagency operations), and capacity building.

Question and Answer Session

Meeting participants expressed concern over the state of the U.S. armed forces, given the emerging challenges, looming budget shortages, tendentious relations among the services, and the poor state of interagency cooperation. One meeting participant asked whether we are running the risk of fighting the last war and whether there is a need to work to create a capability for "counterinsurgency" and nation-building within civilian agencies of the government. This attendee emphasized the need to strengthen the foreign-policy tools of the U.S. government as a whole. We need a better understanding of what civilian tools are available within the government.

A second question focused on the notion of a new division of labor and how the DoD should best prepare to cope with a budget shortfall. The conferee considered the argument for building a single adaptable force against the case for a divided force. A panelist noted that current Army configurations do not have true full-spectrum capability and that specialization within ground forces is both feasible and desirable.

Closing Remarks

> How do we fight two wars and prepare for the future? How much of the world's glue are
> we going to be?
>
> —Senior Military Officer

The final speaker said that, while the U.S. government cannot lose its focus on Iraq and Afghanistan, it needs to prepare now for the challenges that remain beyond these two commitments. A new administration will take office in January 2009, and DoD needs to be prepared to help it meet the country's needs. The speaker argued that DoD needs to (1) develop an overall military strategy for the Middle East; (2) reconstitute its forces, which have been stressed by recent operations; and (3) address the issue of what is at risk elsewhere in the world because of the commitments to Iraq and Afghanistan.

A strategy for the Middle East must look at Iran's future role in the region and the necessity of establishing greater stability among Israel, Syria, Lebanon, and the Palestinian territories. While DoD will play a significant role in shaping stability across the region, the whole government requires a framework for a coordinated approach.

Second, the armed forces are stressed by long rotations; the troops are dedicated, but they "need some oxygen," and this includes both individuals and their families. The ground force, particularly the Army, is the center of gravity for the military right now.

Third, the United States, as a global power, needs to consider what risks it faces in the rest of the world. Given the press of constrained resources and the possibility of negative domestic political reactions to the war in Iraq, it is unclear the extent to which the United States will be engaged in the world beyond its borders in the future. Other countries appreciate U.S. leadership and understand that it makes a difference, but continued U.S. engagement raises a number of questions: What does it mean to be globally committed? Should there be a floor in defense spending as a percentage of GDP? How can the United States remain prepared to fight two wars and, at the same time, prepare for the future? What role will the United States play in holding the world together?

The speaker raised a number of other questions that U.S. defense planners must face, from the meaning of deterrence today to the best way to handle the prospect of a terrorist group gaining nuclear weapons. He also stressed the importance of understanding the perspectives of U.S. security partners, many of which face conflicting domestic and regional pressures. The speaker reinforced one of the dominant themes of the conference: the need for better integration of military and civilian efforts, especially in Iraq.

Question and Answer Session

Responding to a question on promoting civilian-military integration, the speaker noted that this is a long-term problem. There are stovepipes and barriers to cooperation throughout government that need to be addressed. In response to audience questions about the possibility of revisiting the debate on roles and missions, the speaker recalled a similar debate that took place in 1994 and 1995, which was largely unproductive. He said that, since then, the services have made strides toward jointness and that he fears such a debate might jeopardize interservice relations. Instead of a broad, open-ended discussion of roles and missions, he suggested focusing on five or six areas that deserve the most attention. Another audience member observed that there are probably four of five areas of overlaps or gaps. Perhaps a private discussion among the service chiefs and the Chairman of the Joint Chiefs of Staff would provide an opportunity for progress.

Conference Agenda

Meeting America's Security Challenges Beyond Iraq

A Conference Sponsored by
RAND Project AIR FORCE and the Center for Naval Analyses

RAND Fourth Floor Conference Center
Tuesday, December 4, 2007

Agenda

0800	Gather, Sign in, Continental Breakfast
0825	Welcome
0830–0930	Session I: Overview of the Emerging Security Environment

- Overview of Trends Affecting U.S. Security Interests
- "Five Long Wars:" Security Challenges in the Coming Decade

0930–1200	Session II: Key Threats and Challenges for the Future

- China's Strategic Challenge in 2015
- Nuclear-Armed Regional Adversaries—How "Deterrable" Are They Likely to Be?
- Terrorism and Insurgency: The Changing Threat
- Iran: A Multidimensional Challenge

1200–1300	Lunch and Midday Presentation
1300–1430	Session III: Priority Capabilities for Securing U.S. Interests

- Creating Global Effects
- Strategy and Capabilities for Countering Terrorist and Insurgent Groups
- Capabilities Required for Countering State Adversaries

1440–1620	Session IV: Fielding the Requisite Forces in a Resource-Constrained Environment

- The Outlook for Defense Spending, FY09–FY20: Toplines and Competing Priorities
- A New Division of Labor Among America's Armed Forces
- U.S. Ground Forces: Options for Future Development

1630–1730	Closing Remarks
1730	Reception

Bibliography

Congressional Budget Office, *Long-Term Implications of Current Defense Plans: Summary Update for Fiscal Year 2008*, Washington, D.C., December 2007.

Grissom, Adam, and David Ochmanek, *Train, Equip, Advise, Assist: The USAF and the Indirect Approach to Countering Terrorist Groups Abroad*, Santa Monica, Calif.: RAND Corporation, 2008. Not available to the general public.

Hoehn, Andrew R., Adam Grissom, David A. Ochmanek, David A. Shlapak, and Alan J. Vick, *A New Division of Labor: Meeting America's Security Challenges Beyond Iraq*, Santa Monica, Calif.: RAND Corporation, MG-499-AF, 2007. As of March 19, 2008:
http://www.rand.org/pubs/monographs/MG499/

Hoffman, Bruce, "The 'Cult of the Insurgent': Its Tactical and Strategic Implications," *Australian Journal of International Affairs*, Vol. 61, No. 3, September 2007, pp. 312–329.

Ochmanek, David, and Lowell H. Schwartz, *The Challenge of Nuclear-Armed Regional Adversaries*, Santa Monica, Calif.: RAND Corporation, MG-671-AF, 2008. As of April 10, 2008:
http://www.rand.org/pubs/monographs/MG671/

Office of the Director of National Intelligence, "Iran: Nuclear Intentions and Capabilities," *National Intelligence Estimate*, Washington, D.C.: National Intelligence Council, November 2007.

Pew Research Center, *Muslim Americans: Middle Class and Mostly Mainstream*, Washington, D.C., May 22, 2007. As of March 19, 2008:
http://pewresearch.org/pubs/483/muslim-americans

U.S. Department of Defense, *Quadrennial Defense Review Report*, Washington, D.C., February 6, 2006.